Fixing Genes and Treating Disease

A Book about Gene Therapy

by Maria Kefalas & Pat Carr
Illustrations by Lela Meunier

Archway Publishing books may be ordered through booksellers or by contacting:

Archway Publishing
1663 Liberty Drive
Bloomington, IN 47403
www.archwaypublishing.com
844-669-3957

Funding for this project was made possible with support from The Calliope Joy Foundation, Orchard Therapeutics and Spark Therapeutics.

ISBN: 978-1-6657-0750-3 (sc)
ISBN: 978-1-6657-0751-0 (e)

Print information available on the last page.

Archway Publishing rev. date: 05/31/2021

This book is dedicated to the
brave families who have sacrificed
so much in pursuit of a miracle.

Message from the authors

On December 19, 2017, the Food and Drug Administration (FDA) made history and granted approval for the first gene therapy to treat an inherited disease. This achievement came after decades of disappointments that had led many to doubt whether gene therapy would ever become a reality.

Thousands of disorders are monogenic, in other words, caused by a single faulty gene; diseases like hemophilia, Tay-Sachs disease, cystic fibrosis, and sickle cell anemia.

We hope this first approval will usher in a new era of hope for millions of patients, particularly children, who will get a chance to attend school and run and play.

So much has been accomplished, but there is still much work to be done.

-Maria Kefalas and Pat Carr

Every human being has around 30,000 genes. In fact, we have two copies of each of those genes, one inherited from our mother, the other, from our father. Our genes tell our cells what proteins to make.

Each protein is a tiny machine, and every cell in our bodies is built out of millions of these little machines working together.

Proteins break down food, transport energy, and keep our cells healthy.

When these genes do not do their jobs, our bodies do not produce enough of a protein. Thousands of serious illnesses are caused by being born with a broken gene. Doctors hoped that one day they would be able to repair these broken genes by replacing them with a working copy.

This kind of medicine is gene therapy.

Scientists first learned how to alter the genes of dozens of species of plants and animals. They created tomato plants that were resistant to cold and tobacco leaves that could glow in the dark.

The same scientists believed they
could use this knowledge about
plants and animals for human beings.

Scientists also needed a way to deliver the repaired genes into our bodies. And they quickly realized that there was already something in nature that did this job: viruses.

Viruses are really good at getting inside our cells and telling them what to do.

But first, doctors needed a way to remove the parts of the virus that make us sick. Viruses stripped of their dangerous parts are called vectors.

Vectors are like ships transporting cargo, and doctors built vectors out of viruses to transport repaired genes into our cells.

In 1990, a 4 year old girl named Ashanti DeSilva became the first person in the world to be treated with gene therapy. Ashanti's body could not produce a single protein her immune system needed to fight off infection and bacteria.

The doctors removed some of Ashanti's blood cells from her bone marrow. Outside the body, the cells had new, repaired copies of the gene inserted. It was like the doctors were infecting Ashanti with the gene her body needed.

And, in time, Ashanti's body started producing cells with the repaired gene on her own. Ashanti got stronger and could return to school and play with her friends in the sun. Ashanti studied music in college. Today, she is 30 years old.

Because of brave patients like Ashanti and dedicated doctors who care for them, we are just starting to learn what gene therapy will make possible.

Printed in the United States
by Baker & Taylor Publisher Services